FOOD SCIENCE

DR. SHALINI SEN

(Ph.D., NET, SLET)

INDEX

TOPIC No.	CONTENTS
Topic 1	Food science and Nutrition
Topic 2	Physical and chemical properties of foods
Topic 3	Evaluation of food quality
Topic 4	Cooking and Processing of food, food preservation
Topic 5	Food Additives and Food pigments
Topic 6	Food standards, Microbiology of foods, HACCP and Food packaging
Topic 7	Type of Menus and Services
Topic 8	New Product Development

Unit I – FOOD SCIENCE AND FOOD SERVICE MANAGEMENT

TOPIC 1- Food science and nutrition

Food science: It can be defined as the application of the basic sciences and engineering to study the fundamental physical, chemical and biochemical nature of foods and the principles of food processing.

Food science covers all aspects of food material production, handling, processing, distribution, marketing and final consumption.

Food technology: The use of the information generated by food science in the selection, preservation, processing, packaging and distribution, as it affects the consumption of safe, nutritious and wholesome food.

Food engineer's deal with the conversion of raw agricultural products such as wheat into more finished food products such as flour or baked goods.

The Education Committee of the **Institute of Food Technologists (IFT)** adopted a undergraduate curriculum in food science.

Courses in Food Science:

- ✓ Food chemistry
- ✓ Food analysis
- ✓ Food microbiology
- ✓ Food processing
- ✓ Food engineering

Nutrition: is the science that deals with digestion, absorption and metabolism of food.

Nutrition is the science that interprets the relationship of food to the functioning of living organism.

Father of Nutrition: "Antoine Lavoisier" ...... In 1770 he developed the concept of metabolism. He also Father of Chemistry.

In 1926, nutrition mad individual subject. That time "Marry Swartz Rose" was professors in nutrition at Columbia University.

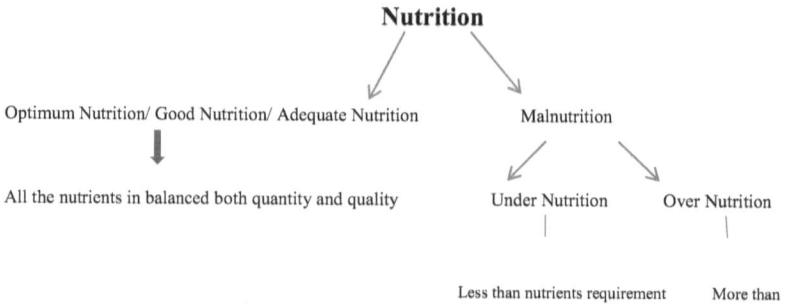

Nutrition

Optimum Nutrition/ Good Nutrition/ Adequate Nutrition

All the nutrients in balanced both quantity and quality

Malnutrition

Under Nutrition   Over Nutrition

Less than nutrients requirement   More than

Health: According to WHO (1948) is a state of complete physical, mental, spiritual and social wellbeing and not merely the absence of disease or infirmity.

TOPIC 2: Properties of food

Properties

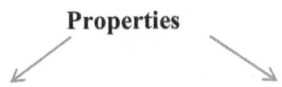

Physical	Chemical

Physical

- Color
- Texture
- Shape
- Weight
- Surface area
- Density
- Appearance
- Volume
- Porosity
- Drag coefficient
- Center of gravity

Chemical

Reactivity
Toxicity
PH
Conductivity
Tarnishing
Fermenting
Oxidation

Difference in Physical and chemical properties

Physical	Chemical
It can be only observed. Identification of foods not change	Change food identity
All changes in state(freeze, melt, condense, evaporate and sublimate)	Develop new product. Changes observed in color, odor, heat etc.
Ex: Boil, dissolve, break, spilt, grind, cut, crush etc.	Ex: Decompose, ferment, oxidized etc.

Topic 3. Quality evaluation of foods

Food Quality: is the ultimate criterion of the desirability of any food product.

Evaluation of Food Quality

Sensory/ Subjective/Organoleptic method   Objective method

   1) Sensory Method: assessed by human sensory organs.

Sensory characteristics of food

i)   **Appearance:** Include foods size, shape, color, structure, transparency or turbidity, dullness or gloss, brightness, completeness of cooking and degree of wholeness or damage.

Example:

- Color of a fruit indicates how ripe it is, and color also is an indication of strength (as in tea or coffee), degree of cooking, freshness, or spoilage. Consumers expect foods to be of a certain color, and if they are not, it is judged to be a quality defect.
- The same is true for size, and one may choose large eggs over small ones or large peaches over small ones.
- Structure is important in baked goods. For example, bread should have many small holes uniformly spread throughout and not one large hole close to the top.

- Turbidity is important in beverages; for example, orange juice is supposed to be cloudy because it contains pulp, but white grape juice should be clear and without any sediment, which would indicate a quality defect.

ii) **Texture:** refers to those qualities of a food that can be felt with the fingers, tongue, palate, or teeth.

Example:

- ✓ Crisp crackers or potato chips
- ✓ crunchy celery
- ✓ hard candy
- ✓ tender steaks
- ✓ chewy chocolate chip cookies
- ✓ Creamy ice cream.

The texture of a food can change as it is stored, for various reasons:

- ✓ If fruits or vegetables lose water during storage, they wilt or lose their turgor pressure, and a crisp apple becomes unacceptable and leathery on the outside.
- ✓ Bread can become hard and stale on storage.
- ✓ Products like ice cream can become gritty due to precipitation of lactose and growth of ice crystals if the freezer temperature is allowed to fluctuate, allowing thawing and refreezing.

iii) **Flavor:** It include odor (aroma), taste, mouth feel.

Taste buds near the tip of tongue: Sweet, Salt

On the side: Sour

Back side: Bitter

iv) **Astringency:** It is dry pucker sensation due to precipitation of the proteins in saliva. Ex: unripe fruits.

v) **Consistency**: Hard, soft, thick, thin….. Temperature may affect the consistency like ghee, butter, cheese, ice cream etc.

vi) **Psychological factors**

Conducting sensory tests:

- **Panel members:**

i) **Trained panel:** Fully trained and experience person. Judge to difference in specific characteristics between stimuli and direction or intensity. Small panel- 5 to 10

ii) **Discriminative, communicative or semi- trained panel:** Technical people and their families, familiar with the qualities of different types of food. Panel member- 25 to 30

iii) **Consumer panel:** Untrained, chosen randomly. Large panel-not less than 100.

- **Testing laboratory:**

i) Reception room

ii) Sample preparation room

iii) Test booths

✓ **Preparation of samples:**

i) Homogenous lot

ii) Same temp.

iii) Optimum level

iv) Kept constant

v) Give sample code

✓ **Testing time:**

i) Between 10 to 12 in the morning

ii) Too many samples should not be given (not more than 4 to 5 sample at a time)

✓ **Evaluation card:**

i) Typed or printed

ii) Simple

iii) Give direction

iv) According to experiment

v) Date and name of judge and food products

Types of tests

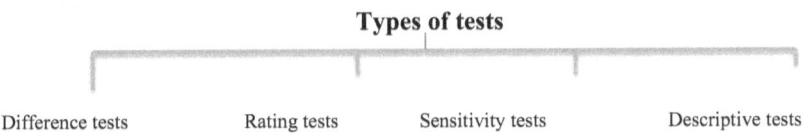

Difference tests   Rating tests   Sensitivity tests   Descriptive tests

A. Difference tests:

✓ **Paired Comparison test:** samples are given in pairs. Both samples may be same or difference. Differences are present in one characteristic like sweetness, bitterness.

✓ **Duo-Trio Test:** Three samples, two identical and one different. Firstly an identical sample which is also known as reference sample R given to the panel. Asked to match one of these with the first. Possible the chance of guess. The chance of probability is one –half.

✓ **Triangle test:** Three samples, two identical and one different. The judge is asked to determine which of the three the odd sample is. All three samples are unknown; the chance of probability is one-third.

B. Rating test:

This test gives more quantitative data than difference tests and can be used for the analysis of more than two samples at the same time.

✓ **Ranking Tests:** This test is used to determine how several samples differ on the basis of a single characteristic. A control need not be identified. Panelists are presented all samples simultaneously (including a standard or control if used) with code numbers and are asked to rank all samples according to the intensity of the specified characteristics.

✓ **Single Sample (Monadic) Test:** This test is useful for testing foods that have an after taste or flavor carry over which precludes testing a second sample at the same session. The panelist is asked to indicate the presence or absence and/or intensity of a particular quality characteristic. With trained panelists, the completed analysis of two or more samples evaluated at different times can be compared. Also, in market and consumer analysis, the results of different samples evaluated at different times by a different set of untrained panelists can be compared.

✓ **Two Samples Difference Test:** This test is a variation of the paired comparison test and measures the amount of difference. Each taster is served four pairs of samples. Each pair consists of an identified reference and coded test sample. In two pairs, the test sample is a duplicate of the reference sample. In the other two pairs, the test sample is the test variable. The panelist is asked to judge each pair independently as to the degree of difference to '3' representing extreme difference. Additional questions on direction of difference can also be asked.

✓ **Multiple Sample Difference Test:** In this test, more than one test variable can be evaluated per session but with reduced reliability. Each panelist is served 3-6 samples depending upon the number of test variable. One sample is a known standard. The panelist compares each coded sample with the known standard. One coded sample is a duplicate of the standard. Direction and degree of difference is also to be judge.

✓ **Hedonic Rating Test:** The hedonic rating test is used to measure the consumer acceptability of food products. From one to four samples are served to the panelist at one session. He is asked to rate the acceptability of the product on a scale, usually of 9 points, ranging from 'like extremely' to dislike extremely'. The results are analyzed for preference with data from large untrained panels.

9 – Like extremely

8 – Like very much

7 – Like moderately

6 – like slightly

5 – Neither like nor dislike

4 – Dislike slightly

3 – Dislike moderately

2 – Dislike very much

1 – Dislike extremely

0 - Reason

✓ **Numerical Scoring Test:** One or more samples are presented to each panelist in random order or according to a statistical design. The panelist evaluates each sample on a specific scale for a particular characteristic indicating the rating of the samples. The panelist is trained to follow the sensory characteristics corresponding to the agreed quality descriptions and scores.

✓ **Composite Scoring Test:** The rating scale is defined so that specific characteristic of a product are rated separately. This method is helpful in grading products and comparison of quality attributes by indicating which characteristic is at fault in a poor product.

C. Sensitivity Tests

Sensitivity tests are done to assess the ability of individual to detect different tastes, odors and feel the presence of specific factors like astringency or hotness (pepper).

✓ **Sensitivity-Threshold Test:** Sensitivity tests to measure the ability of an individual to smell, taste or feel specific characteristics in food.
There are mainly three types of threshold as described below
a. Stimulus detection
b. Recognition identification
c. Terminal saturation
✓ **Dilution Test:** Establish the smallest amount of an unknown material, developed as a substitute for a standard product that can be detected when it is mixed with the standard product, e.g., margarine in butter, dried whole milk in fresh milk, synthetic orange flavor ingredients with natural flavor and so on.

D. Descriptive Flavor Profile Method

This is both qualitative and quantitative description method for flavor analysis in product containing different tastes and odor e.g. tomato ketchup.

Limitations of sensory evaluation

1 Highly variable

2 People with colds or other health problems

3 Emotional burdens

2. OBJECTIVE EVALUATION METHOD

Methods of evaluating food quality that depend on some measure other than the human senses.

Advantages:

- Confidence
- Accurate
- Less subjected to errors
- Permanent record
- Emotional burdens and individual ability can be overcome.

Disadvantage:

- Time consuming
- Expensive
- Technical knowledge is required
- Instruments may not be available sometimes
- Some aspects of food cannot be evaluated by objective methods e.g., flavor.

Tests Used for Objective Evaluation:

Chemical methods: Chemicals are estimated in food spoilage like peroxides in fats. Adulterants in food e.g., presence of starch in milk, metanil yellow in turmeric powder and loss of nutrients during cooking can be estimated.

Physico-chemical methods:

a. Measurement of hydrogen ion concentration can be found by the use of **PH meter.**

b. **Digital salt meter:** It indicates salinity percentage.

c. Sugar concentration can be found by **Refractometer**. Light is refracted as it passes through sugar solution, with the specific values being calibrated in degrees, **Brix,** an indication of the percent of sucrose in the solution.

   Brix or Balling hydrometer gives directly the percentage of sugar by weight in the syrup. Brix is defined as percent sucrose measured by a Brix hydrometer.

d. **Polaris cope** is used for quantitative analysis of sugar.

e. **Butyrometer**: Measuring the butter content of milk. Milk is mixed with certain volume of ether which dissolves the butter. Then an equal volume of alcohol is added. The butter floats on the surface.

Microscopic examination method: use microscope and used in

1. Idli batter
2. Examination of starch
3. Spoilage of the food
4. Size of crystals in sugar is related to smoothness of the product.
5. Number and size of the air cells in batters and foams.

Physical methods:

a. **Weight:** Weight of food indicates the quality like in case of apple or egg.

b. **Volume:** Measured by using measuring cups (liquid). Solid food volume can be found by displacement method. In this method the volume can be calculated by subtracting the volume of seeds held by a container with a baked product from that of volume of seeds without the baked product. Usually mustard seeds are used.

c. **Specific volume:** Measurement of bulk volume in a porous and spongy product like idli is difficult. The volume may be measured by displacement with solvents like kerosene. The idli is given a momentary dip in molten wax to seal off the pores. Increase in volume is taken as the measure of its bulk volume.

Specific volume = Bulk volume/Wt. of the substance

d. **Index to volume:** Measuring the area of a slice of food with a planimeter. First tracing detailed outline of a cross section of the food. Can be done with a sharply pointed pencil or a pen or by making a clear ink blot of the cross section.

e. **Specific gravity:** It is measure of the relative density of a substance in relation to that of water. The measurement is obtained by weighing a given volume of the sample and then dividing that weight by the same volume of water. Used for comparing the lightness of products e.g., egg white foams.

f. **Moisture:**

- **Press fluids**: Initial weight of the sample is noted. After the appropriate pressure has been applied for a controlled length of time, the sample is again weighed e.g., juiciness of meats, poultry and fish.

- **Drying:** The weight of the original sample is determined and then the food is dried until the weight remains constant.

   Moisture content = initial – dried weight / initial weight × 100 = %

- **Karl Fischer Titration:** In 1990 Karl Fischer showed that food to be analyzed by this method is homogenized in a high-speed blender at speeds up to 7,500 rpm to release the water and the water is titrated with Karl Fischer reagent until all the water has reacted with the reagent.

g. **Wettability:** Baked products can be tested for moisture level by conducting a test for wettability. The sample is weighed before being placed for 5 seconds in a dish of water. Immediately at the end of the lapsed time, the sample is removed from the water and weighed again to determine the weight gain.

h. **Cell structure:** Cell structure of baked products is an important characteristic to measure the uniformity, size and thickness of cell walls.

- Photocopies of cross-section slices
- Size of the grain: This can be found by using photography or ink prints with stamp pad or sand retention e.g. idli.
- Photography: This may be color or black and white. They may not represent the sample size.

i. **Measurement of color:** Color is the first quality attribute a consumer perceives in food. Change of color is generally accompanied by flavor changes.

- Color Dictionaries: The dictionary of Maerz and Paul is most commonly used. The dictionary consists of 56 charts. Seven main groups of hues are presented in order of their spectra. For each group there are 8 plates.
- Disc colorimeter: The disc has radial slits so that a number of them may be slipped together with varying portions of each showing. The discs are spun on a spindle at about 2700 rpm so that the colors merge into a single hue without flickering.

- Colored chips: A sample method is to match the color of the food with the color chips or color glass, chart or color tiles.
- Spectrophotometer: Tube with the liquid is placed in a slot and light of selected wavelength is passes through the tube. This light will be differentially absorbed depending upon the color of the liquid and the intensity of the color.

INSTRUMENTS USED FOR TEXTURE EVALUATION

Rheology is defined as the science of deformation and flow of matter. It has three aspects – **elasticity, viscous flow and plastic flow.**

Rheological properties of foods –

- To determine the flow properties of liquid food stuffs.
- To ascertain the mechanical behavior of solids foods when consumed and during processing.

Instruments used for Liquids and Semisolids:

The resistance or internal friction to the flow of liquids is normally known as viscosity. Viscosity or consistency is an important Factor in influencing the quality bb of a large number of food products. Creams style corn, salad creams, tomato products, jellies, jams, mayonnaise, syrups, and fruit pulps.

Percent Sag: The depth of a sample such as jelly is measured in its container by using a probe. The product then is unmolded onto a flat plate. The greater the percent sag, the tenderer is the gel.

Percent Sag = depth in container – depth in plate /depth in container × 100

Stormer viscometer: It is used measure the viscosity or consistency of certain food products and to give an index of the resistance of the sample to flow. The number of seconds required for the rotor to make 100 revolutions has been used to measure the consistency of some food samples.

Brookfield Synchrolectric Viscometer: This is based on measurement of resistance to rotation of a spindle immersed in the test material. Measuring the consistency of custards, pie fillings, tomato products, cream style corn, mayonnaise, salad dressings and dairy products.

Bostwick Consistometer: It useful for measuring the consistency of tomato ketchup and sauce. It consist a channel (2×12") with sides which are 2" high. This instrument is based on the theory that the length of flow is proportioned to consistency.

Efflux- Tube Viscometer: It measures the time necessary for a quantity of fluid to pass through an orifice or capillary under standard pressure e.g., tomato puree.

Adams Consist meter: It measures the unrestrained flow in all directions by means of concentric circles. A simple version of this principle is used in Line spread test.

Penetrometer: It measure tenderness of some foods. This device consists of a plunger equipped with a needle or cone that is allowed to penetrate the sample by gravitational force for a selected period of time. The larger the reading the longer the distance the more tender is the product, e.g., Gels, baked products. The bloom gelometer is a special type of penetrometer.

Brabender farinograph: It measures the plasticity of wheat dough for preparing bread products. Study the physical properties of the dough by recording the force required to turn the mixer plates through the dough. The force required increases as the solution develops during mixing and later decreases as solution is slowly broken down by over mixing.

Instruments used for solids: Food texture can be reduced to measurements of resistance to force.

> ➤ If we squeeze food that it remains as one piece this is called **compression.** e.g. bread.
> ➤ If we apply a force so that one part of the food slides past another it is **shearing.** E.g. chewing gum.
> ➤ If the force goes through the food so as to divide it as we call it **cutting.** e.g. cutting an apple.
> ➤ If the force is applied away from the material, the food pulls apart by which we measure **tensile strength.** e.g. chapati

Magness- Taylor pressure tester (compression): It consists of a plunger of variable diameter which is pressed into the fruit to a given depth. The sprint attached to the plunger contracts and measures the compression force, e.g. peas (suitability of peas for the harvest or to find out the correct stage of ripening of a food).

Succulometer (compression): This instrument is used to measure the maturity of corn and storage quality of apples as determined by the volume of juice extracted under controlled conditions of pressure and time.

Tenderometer (compression and shearing): This is an example of an instruments based on shearing force in which compression is preceded by shearing action, e.g. suitability of peas for preservation.

Fibrometer: This is based on the cutting principle and used to differentiate mature stocks from the tender stocks, e.g. green beans.

Shortometer: This device consists of a platform containing two parallel, dull blades on which the sample rests. A third blade is actuated by a motor to press down on the sample until the sample snaps. The force required to break the sample is the measure of the tenderness of the product. Find out the breaking strength of idli, pastries and cookies.

Cristal texturometer (cutting): This is designed with series of rods which are pushed into the meat sample. The harder the meat more force is required to penetrate.

Voldokevich bite tenderometer (cutting and shearing): Imitate the action of teeth on food. It records the force of biting on a piece of food which results in deformation and this determines the total energy utilized for this deformation, e.g. meat and meat products.

Grinding and extensibility: The power used by a household food grinder is measured. Increased toughness would increase the current consumption of the grinder. Extensibility has proved to be inversely related to tenderness.

Kramer shear press: This is a multipurpose instrument with same power unit and with different fest cell assemblies. This instrument is widely used.

Tensile strength: An instrument used to find out the tensile strength of chapathi.

Compressimeter: Related to the shear press.

Warmer – Bratzier Shear: Measure tenderness of meat

Shear press

Universal Testing machine: It measure cohesiveness, adhesiveness, hardness, springiness, gumminess, chewiness and fructurability.

Topic 4: Effects of cooking and processing

Preliminary preparation of foods:

- **Peeling and Stringing**: Peeling and stringing are the method, which refers toward the removal of non-edible portions of fruits and vegetables.
- **Cutting and grating:**
- **Cut:** To shape with knife or dived into pieces
- **Mince:** Refer to the very chop
- **Chop**: Refer to the cut into no specific shape
- **Slice:** Refer to the cut into uniform slice/pieces
- **Dice:** Refer to the cut into small uniform cubes
- **Grate:** Refer to the cut finely with a grater
- **Sieving:** Sieving refers to the removal of coarse fibers and insects. It is also done in preparing cakes for blending of flour with baking powder.
- **Soaking:** Soaking is done sometimes with plain water or sometimes with salted water.
- **Coating:** Coating refers to the covering a food layer with the help of bread crumbs, flour before cooking it to make crispy.

Different methods of coating involves-

- **Dredging:** Substance used in coating such as flour. Powdered-bread crumbs etc. It is a method in, which, food is, coated with a fine dry powder Substance before cooking or frying the food.
- **Breeding:** 3 steps

I) Food product dredged with flour

II) Dipped the food product in liquid

III) Coat with the crumbs

- **Battering:** Batter is semiliquid in texture generally it consists of egg liquid mixture thickened with flour rice, creamed to make smooth consistency. Food product into the batter.
- **Blanching:** It involve 2 steps
    - i) Water plunging food into boiling liquid
    - ii) Immersing in cold water

Advantages of blanching:

- ✓ Easy to peel
- ✓ Caning and freezing is preliminary method.
- ✓ Help in removal of micro organism
- ✓ Improve the color
- ✓ Destroy enzymes present in food.
- **Marinating:** Soaking of food in marinade mixture before cooking to add flavor in it food can be marinating according to taste and requirement of the cooking recipe.
- Sprouting and Germination
- Fermentation
- Roasting
- Drying
- Filtering

Cooking Methods:

    I)   **Moist heat method:**

        i)   **Boiling:** Refers to any liquid, which is bubbling and break down rapidly food that are cooked. Excess amount of water is use to do boiling at 100°C.

Advantages: Simple method, uniform cooking, special skill not required, help in removal of soluble starch.

Disadvantages: Time consuming method, loss of flavor and texture of food, loose water soluble nutrients.

        ii)   **Simmering:** It refers to the food cooking in a pan with well-fitting lid at temp. 82 to 99° C below the boiling point.

Advantages: Prevention of scorching and burning, minimum losses due to leaching.

Disadvantage: More fuel requires, long period of cooking there is loss of sensitive nutrients.

        iii)   **Poaching:** In this method 80- 85°C temperature below the boiling point along with minimum amount of liquid is required for cooking. Ex: egg, fish, fruits etc.

Advantages: Quick method, no fat used, easily digestible.

Disadvantages: Water soluble nutrients may be leached in water, bland in taste

iv) **Stewing:** Cooking using small quantities of water in a pan with tight lid. Ex: meat, vegetables, legumes, stock.

Advantages: Flavor are retained, loss of nutrients does not take place.

Disadvantages: Time consuming method

v) **Steaming:** Steamer is used and water should be boiled at 100°C before food is placed in the steamer. They have 3 types:

a) **Wet steaming:** Steam is in direct contact with food, e.g., idli

b) **Dry steaming:** Double boiler techniques are used for cooking, e.g., melting a chocolate.

c) **Water less cooking:** Steam is originated from the food itself. Aluminum foil is used to wrap the food & food is cooked by its own steam.

vi) **Pressure cooking:** Reduce cooking time steam is trapped & kept under pressure, so that steam can be raised 100°c.

Dry heat method (Air as medium of cooking)

- **Grilling:** Red hot surface or heated by radiation are the method by which food is cooked.

Food in placed between, below & above the red hot surface or sometimes under the heater until browning takes place. Ex: - Barbecues, papad, brinjal, sweet potato, phulkas etc.

- **Pan broiling or Roasting:** Refers to uncover cooking using a heated metal or a frying pan, e.g., Roasting rava, groundnut etc.
- **Baking:** Hot air is req. for baking. When hot air is combined with steam it help the food to bake &it looks brown, crisp, and soft in texture.
    - Tandoor: Chicken, mutton, tandoori nan
    - Oven: Electric oven and gas oven

Dry heat method (Fat as medium):

- **Sautéing:** Once the pan is greased, foods are occasionally tossed, so that all pieces come in contact. Ex: - veg.
- **Shallow fat frying:** Ex: - Parantha, cutlets, tikkas food become crispy brown outside, soft & tender inside.
- **Deep fat frying:** Uniform cooking ex: - Pakoda, mathri, kachori etc. When repeatedly oil is used, large amount of oil or fats is absorbed.

Combination of cooking method:

- **Braising:** Refers to the combination of method such – roasting + stewing

Effect of cooking on nutritive value:

i) **Cereals**
- **Dextrinisation:** It refers to non-enzymatic browning. Ex: - Parantha, mathri, kachori, samosa etc.

    Starch →Dry heat → Dextrin (brown color)

Dextrin reduces the thickening power of starches.

- **Gelatinization: Starch → moist heat → gel like structure**. Ex: - kheer, khichdi etc.
- **Retro gradation:** Means starch paste become less soluble after cooling.
- **Syneresis:** Water is being pushed out of the gel, due to tightly packed amylose molecules.

ii)   Pulses

- The digestibility & availability of amino acids is enhanced / increased on cooking as anti-nutritional factors like trypsin inhibitors etc. are destroyed.
- Quality of protein is reduced on excessive heating.
- Methionine amino acid is lost on heating.
- Moist heat improves the protein quality of pulses.
- Dry heat method as available lysine is decreased in roasted pulses in comparison to boiled & pressure cooked pulses.

iii)   Milk

- Coagulation of milk protein, casein does not take place on heating.
- Flocculants precipitates that settles on the sides of the containers, is formed by the action of heat on albumin present in milk.
- Losses iodine contents on heating.

iv)   Eggs

- Denaturing of **Avidin** takes place on heating, thus improving the availability of biotin in egg white.

- Some loss of thiamin & riboflavin.

Food Processing

i) **Rice**

➤ **Milling:** Cleaning of the paddy is done to remove the impurities. It is done by two methods – **Home pounding & mechanized rice mills**

Rancidity develops more quickly in home pounding

Three forms of rice is available after processing –

- **Smaller size – brewers rice**
- **Medium size – screening**
- **Larger size – second heads**

➤ **Parboiling:** It involves steeping the paddy in cold water for a few days boiling it till the grain is soft. Husk is more easily detached. Nutritive value is increased by the improved protein efficiency ratio, digestibility also increased. Excellent source of B- complex vitamins.

ii) **Pulse**

➤ **Soaking**

➤ **Decortication:** The ingestible part which is the form of fibrous seed coat is removed.

➤ **Fermentation**

➤ **Germination**

iii)   Milk

➢ **Clarification:** Remove out the dirt, filth & bacteria from the milk. The pasteurization is the next step after clarification.

➢ **Pasteurization:**

- High temp. Short time method **(HTST)**: 71.7° for 15 second.
- Low temp. Long time **(LTH) or holding method**: 62.8°c for 30 min.
- Ultra pasteurized method **(UHT)** or Ultra high temp: 137.8°c for 2 second.

➢ **Homogenization:** Stable emulsion of milk fat & milk serum is made by mechanical treatment & the process is termed as homogenization.

iv)   Fruit

➢ **Canning:** Sorting & grading of fruits → washing of fruits →peeling & coring of fruit →can filling of fruit →syruping of fruit with hot sugar →lidding of cans →sealing of cans

➢ **Drying**

➢ **Freezing**

v)   Vegetables

➢ **Canning**

➢ **Freezing**

➢ **Drying**

➢ **Pickling:** This type of processing preserves veg. in vinegar or a salt solution (brine) or a combination of the two.

➢ **Pureeing**

➢ **Sulfuring**

Food Preservation

<u>Principles of food preservation</u>

1. Preservation by microbial decomposition:-

- By keeping out micro-organisms (Asepsis)

- By removal of micro-organisms (Filtration)

- By hindering the growth & activity of micro-organisms, e.g., low temp., drying, anaerobic condition.

- By killing the micro-organisms

2. Prevention of self-decomposition of food:-

- By destruction or inactivation of food enzymes

- By delay of chemical reaction

3. Prevention of damage caused by insects, animals & mechanical causes

<u>Methods of Food Preservation</u>

1. By low temperature:-

i) Freezing

- **Slow-freezing process:** At **-4°C to -29°C** temperature food are placed in refrigerated. It also called **sharp freezing**. Home freezing is done by Sharp method, which require freezing from 3 to 72 hrs. **-15 to -21°C** temperature for fruits &vegetables.

- **Quick-freezing process:** Large quantity of food can be frozen in short duration.
- **Dehydro freezing:** Fruits & vegetables are dried about 50% of its original wt. & volume. Then freezing the food to preserve it.

2. By High Temperature:-
- Pasteurization
- Heating up to 100°C
- Heating above 100°C

3. Preservation by Osmotic pressure:-
- High osmotic pressure inhibits microbial growth
- By high concentration of sugar
- By high concentration of salt

4. By drying procedure:-
- **Freezing Drying:** Refers to the removal of water from a product while, it is frozen by sublimation. Food is placed in a vacuum chamber & a small amount of heat is applied reducing the moisture.
- **Sun drying**
- **Drying by Osmotic**
- **Spray drying:** In spray dries milk & eggs are dried to a powder, in which the liquid is atomized & sprayed into a hot air steam for air instant drying.

5. Food preservation:-

- **Sorbic acid:** Inhibits the growth of yeasts & molds, e.g., yoghurt sweets, soft drinks, frozen pizzas, desserts, and fillings.

- **Benzoic acid:** Antibacterial & antifungal, e.g. - jam, jellies, desserts, juice, marmalades, fruit yoghurt, soft drinks.

- **Propyl 4-hydroy benzoate:** Antimicrobial, e.g. dessert, sauces, fruit pulp, purees, prickles.

- **Sulphur-dioxide-** Inhibits growth of fungal & non-enzymatic browning. Ex. - fruits & vegetables products, soft drinks, beers, and sausages dehydrated vegetables.

   Related preservatives- Sodium sulphite, sodium metabisulphite, potassium meta-bisulphite (KMS).

- **Nisin:** Antibacterial ex.- cheese
- **Sodium nitrite:** inhibits growth of clostridium botulin, ex:- processed meat.
- **Sodium nitrate:** Curing salt, color retention, ex.- processed meat
- **Acetic acid:** Antibacterial, ex. - pickles, chutney, cheese, sauces.
- **Propionic acid:** Anti-fungal. Ex. - baking, dairy products, pizzas, processed cheese.

Unit 5- FOOD ADDITIVIES

Defined as a substance or mixture of substances, other than a base foodstuff, which is present in a food as a result of any aspect of production, processing, storage or packing.

This definition includes both **intentional & unintentional additives**

1. Anti-caking agents: - Prevent lumping & caking by absorbing moisture.

Ex- Table salt, instant mix.

2. Antimicrobial agents: - Prevent the growth of bacteria, molds, fungi & yeast.

Benzoic acid- Ex- Squashes crushes

3. Antioxidants: - Prevent flavor & color changes & retard rancidity & deterioration from exposure to oxygen.

- Lecithin, vitamin C & tocophenols, BHA (Butylated Hydroxy Anisole)

Ex- Dry mix, ghee, butter, edible oil, fats.

4. Colors: - Annatto, carotene, cochineal, chlorophyll, nitrates,

Ex- Ice-cream, biscuits, cakes, sweets etc.

5. Curing & pickling agents: - Impart unique flavor or color to a food, increase shelf-life and stability.

- Sodium nitrite,

Ex - Meat

6. Emulsifiers: - Prevent the separation of oil & water. Provide surface wetting, lubrication & viscosity change.

- Use as- Ammonium phosphatides, lecithin, sorbitous

Ex- chocolate making, cheese making.

6. Firming agents: - To retain texture of canned fruit & veg.

– CaCo3, sodium aluminum sulphate, calcium citrate,

Ex- canned tomatoes

8. Flavor enhances: - MSG (Monosodium glutamate up to 1%) and yeast,

Ex- masala/spices used in noodles/Chinese cookery.

9. Enzymes: - Pepsin, rennet (renin), papain, amylose, pectinase,

Ex - cheese, tenderizing meat, beverage clarifying.

10. Neutralizing: - To remove excess acidity, ex - wine, ice- cream.

11. Stabilizing agents: -Caragean gums, gum arabi

Ex - chocolate, ice-cream, foam stabilizer in beer

12. Humectant: - Prevent undesirable drying of foods & to maintain the moisture level. Glycerin, sorbitol, mannitol, dextrose.

Ex - shredded coconut

13. Sequestrant (chelating agents):- Prevent the deterioration of food due to free metallic ion.

- EDTA, KH2Po4, tartaric acid,

Ex - Soft drinks industries, malted beverages

14. Flour improvers: - For bleaching purpose & dough toughening

- Benzoyl peroxide, Chlorine dioxide, Nitrogen dioxide, Dinitrogen tetraoxide, Potassium bromate etc.

Ex - Flour

15. Leavening agents: - To increase in volume of dough or batter resulting in light fluffy, spongy.

 Baking powder, sodium bicarbonate with another acid component like tartaric acid or calcium hydrogen phosphate.

Ex - Bread, cakes, khaman etc.

16. Sweeteners: - Non-nutritive & Nutritive

17. Clouding agents: - To produce haze in liquid foods.

18. Clarifying agents: - To remove cloudiness – Bentonite, tannic acid, gelatin

Ex- Fruits juices, beer, wines.

FOOD PIGMENT

Six types or classes of food pigments:-

1. Heme pigment: Myoglobin – present in muscles

Hemoglobin – present in blood

Meat →Cut → Purplish → O2 → Red   cooks → Brown

When cook the meat, protein is denatured & heme convent to nicotinamide hemi chrome & they give brown color.

2. Chlorophylls: Present in green leafy vegetables, green veg., and unripe fruit. Not stable pigment

Chlorophylls → Acid processing → Bright green → Dull olive brown pheophytins

Freezing storage – Retain color

3. Carotenoids: Orange, yellow, green colors provide.

Types:-

- Carotene - Carrot, egg yolk, orange
- Xanthophyll – Veg., egg, chicken fat
- Zeaxanthin – Yellow corn, egg, liver
- Cryptoxanthin – Egg, yellow corn, orange
- Physalien – Asparagus, berries
- Bixin – Annatto seeds
- Lycopene – tomato, pink grapefruit, palm oil

- Capsanthin – Paprika
- Astaxanthin – Lobster, shrimp, salmon
- Torularhodin – Rhodotorula yeast
- Canthaxanthin – Mushrooms
- F-Apo-8'- caratenal – Spinach, orange
- Carotenoids elestroy in storage – 20 to30%

4. Flavonoids-

- Anthocyanins – Kyanos – blue

    Anthos – pigments of flowers

    -Anthocyanin's – Apple, black berry, black currant, blue berry
- Flavones
- Flavanones – citrus fruit
- Flavonals – Ouescentin (Apples, grapes)
- Isoflavones – soybean

5. Betalains – Water soluble

Ex- Red – violet – Beta cyanins,

Yellow – Betaxanthins

(present in red beet/beet root)

6. Others – Quinone – walnuts

Xan-thones (mangiferin) – Mango

Tannins – tea, coffee

Water soluble pigment:-

- **Flavanoids**
  - ✓ Anthocyanin – Red to purple (brinjal, pomegranates, blue berries, cherries),
  - ✓ Anthoxanthins – (colorless to pale yellow, green colors)

  Ex- Onions, cauliflowers, GLVs
- Betalins – beet root

Water insoluble pigments:-

- Chlorophyll
- Carotenoids

Natural colors:-

- **Anthocyanin** – Blue to radish

- **Annatto** – i) Bixin – Orange – Ex- butter, cake, popcorn oil,
- ii)   Norbixin – Ex- cheese (cheddar) flavonoid milk & drinks, bakery &confectionary.

- **Cochineal extract** – insect – magenta – red

  Ex- Cakes, alcoholic drinks, beverages, ice-cream, candy, sweets.

- **Luetin** (Antioxidant) – Marigold flower. Give light yellow to intensely yellow color
  Yellow color
- Paprika – red pepper (Give bright orange to red – orange)

Topic 6 -<u>FOOD STANDARD</u>

Food standard agency - National office → Scotland (April, 2015)

Formed – 1 April 2000

Head quarter – France, London

International food standard set by FAO and WHO

Food Standard in India – 1. Compulsory

2. Voluntary

1. Compulsory:

- PFA (Prevention of food Adulteration Act, 1955), Food Adulteration act, 1954.
- Essential Commodities Act, 1954
- FPO (Fruit product order, 1955)
- MPO (Meat product order, 1973)
- Milk &Milk products order include in FSSAI, 1992
- Solvent extracted oils, Flour control order, Vegetable products order, 1967
- Standards on weights & measures, 1971
- Vegetable oil products order, 1947
- Edible oils packaging order- 1988

2. Voluntary:

- AGMARK (Agriculture produce grading & marketing Act,1973)

  Total 222 commodities ex- Cereals, its products, pulses, spices etc.

4 types of grading – 1. Special

　　　　　　　　2. Good

　　　　　　　　3. Fair

　　　　　　　　4. Ordinary

Head office – Faridabad (Haryana)

Central Agmark lab: - Nagpur

- BIS (Bureau of Indian standard):- ISI (1947) is now called BIS (1986).
- FSSAI – Food Safety and standard Authority of India

  Food safety and standard Act, 2006

It standard of food articles & regulate to their manufacture, storage, distribution, sale, import.

Head office – Delhi

HACCP (Hazard Analysis of Critical Control Point)

It established for food safety HACCP system – used all stages – from food production, preparation, processing, packaging, distribution etc.

Run by – FDA (food & Drug administration)

USDA (United States Dep. Of Agriculture)

Principles:- 7

1. Conduct a hazard analysis: Determine food safety hazard & identify the preventive measures.

2. Identify critical control points (CCP): It is a point, step or procedure in a food manufacturing process at which control can be applied & food safety hazard can be prevented, eliminated or reduced.

3. Establish critical limits for each critical control point.

4. Establish critical control point monitoring req.

5. Establish corrective actions

6. Establish procedures for ensuring the HACCP system is working as intended validation ensure that the plants do what they were designed to do, that is; they are successful in ensuring the production of a safe product.

7. Establish record keeping procedures.

HACCP principles included in ISO 22000 (2011).

- Application of HACCP – Fish & fishery products
    - Fresh cut procedure
    - Juice & nectar products
    - Food Outlets

- Meat & poultry products

- School food & service

MICROBIOLOGY

HISTORY:-

- **1810** – Introduced canning for preservation of food by "Nicolas Appert"
- **1813** – "Camble & Donkin" introduce the practice of food processing & use of Sulphur dioxide as meat preservative.
- **1825** – E Dagett was granted for preserving food in tin canes
- **1840** – Fish & fruits were first canned
- **1841** – Freezing of food by "H Benjadelmin"
- **1855** – Grimwade first produce powder milk.
- **1857** – Milk was designated as the transmitter of typhoid by W Taylor.

Food Born disease caused by Pathogenic Organisms:-

- **Bacteria:**

a) Basillus cerus →Cereal product → nausea, vomiting, abdominal pain

b) Clostridium botulinum toxins → Defectively processed meat & fish → paralysis, death due to respiratory failure

c) Streptococcus pyogenes → foods kept exposed or stale in unhygienic surroundings which lead to scarlet fever, septic sore throat.

d) Staphylococcus aureus → foods kept exposed which leads to salivation, vomiting, abdominal pain & diarrhea.

e) Shigalla sonnei → Foods kept exposed which leads to bacterial dysentery.

f) Clostridium perfringens salmonella →Defectively processed & precooked food & raw vegetables grown on sewage which leads to nausea, diarrhea & abdominal pain

g) Salmonella → Enteric fever

h) Protozoa→ Amoebic dysentery

- **Fungal:**

a) Aspergillus flavus → Corn & groundnut which causes liver damage & cancer

b) Claviceps purpurea → Rye & pearl millet which leads to ergotism

c) Penicillium islandicum → Rice which leads to liver damage.

- **Parasitic:**

a) Trichinella spiralis → pork & its products → nausea, vomiting, diarrhea, muscular pain

b) Entamoeba histolytica & Ancylostoma duodenale → Raw veg. grown on sewage which leads to epigastric pain, loss of flood & anemia.

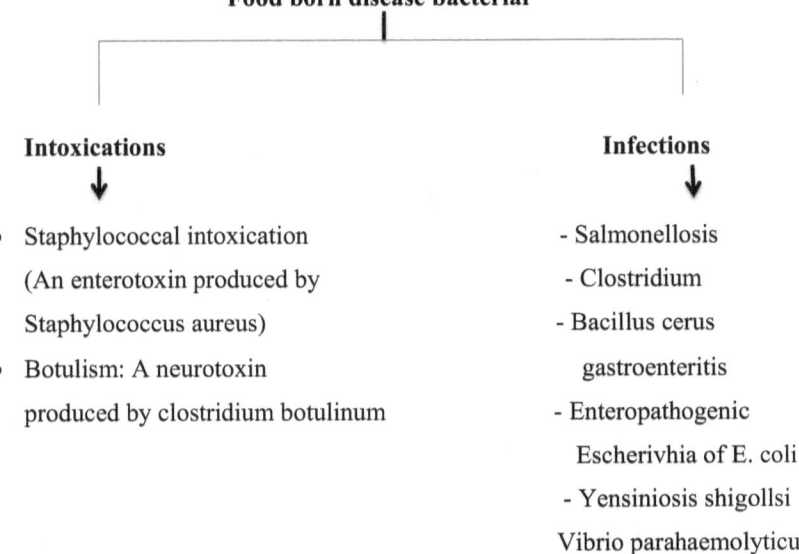

Food born disease bacterial

Intoxications

- Staphylococcal intoxication
  (An enterotoxin produced by
  Staphylococcus aureus)
- Botulism: A neurotoxin
  produced by clostridium botulinum

Infections

- Salmonellosis
- Clostridium
- Bacillus cerus
  gastroenteritis
- Enteropathogenic
  Escherivhia of E. coli
- Yensiniosis shigollsi
  Vibrio parahaemolyticus

Mycotoxin (fungal) (Moulds)

- Aflatoxins → Barley, corn, cottonseed, millet, oats, peanut, rice, soya bean
- Patulin → Apple sap, apple juice, apple cider
- Ochratoxin A → Corn, wheat, barley, white bean, peanuts, egg
- Luteoskyrin → Mold, rice flour
- Sterigmatocystin → Wheat, oats

- Penicillic acid → Dried beans, tobacco
- Alimentary toxic Aleukia (ATA) → Grain
- Roquefortin → Blue cheese, stilton cheese

FOOD PACKAGING

"A packaging provides protection, tampering, resistance and special physical, chemical or biological needs.

Types of packaging:-
➤ Glass:
Testing of glass containers:
- Visual defects
- Critical defects: Cracked or broken glass, choked bore, internal fins, over press etc.
- Major defects: Oil marks, blisters, crizzle, deformation etc.
- Hydrostatic pressure test: Bottles are filled with water & pressure is applied by pump.
- Chemical durability test of glasses

Making of glass bottles

- **Hot end process:** Calcium oxide, lime, silica, soda – lime glass & small amt. of aluminum oxide, ferric oxide, barium oxide, sulphur trioxide & magnesia for about 5 percent of soda – lime glass.
- Before melting, cullet (recycled glass) is added to the stock.
- After stock has been fed into furnace temp. is increased to 1675°F
- One of two method forming methods is applied press and blow or blow and blow.
- **Annealing:** Formation is complete; some bottles may suffer from stress as a result of unequal cooling rates. An annealing oven can be used to reheat & cool glass containers.
- **Cold end process**

- **Closures: -** It is a devices & techniques used to close or seal a bottle, jug, jar, tube etc. It can be a cap, cover, lid, plug etc.

Types of closures:-

- Metal closures – they have EOE (easy opening ends)
- Crown caps – Used in beverage bottles
- Lug caps – They seal the vaccum inside the headspace of the glass bottle, they req. puncturing to open it. It also known as twist – on twist – off caps. Used in jam bottles etc.
- Plastic closures
- Corkers – Use to close wine bottles

- Plastic hinge – open – snap – shut closures: used for liquid products, cooking oils, sauces, fruit toppings.

➢ **Cans**

Types:-

- Full top cans – packing the acidic foods
- Slip over lid – Airtight & temper to sealing
- Coffee cans – Plastic lid convenient to open
- Colapse cans – Bulkiness & transportation can be achieved at low cost by delivering fatten can bodies. Reformed & prior to filling.
- Aerosol cans – Pressurized cans i) Foam

  ii) Wax spray
- Foil laminated fiber cans – Ex – fruit juice, juice concentrates.

➢ **Plastics:**

- Polythene – Boil in the bag product
- Polyamide (nylon) – Especially for food containing fat
- Type 1 – PET (water container)

  APET (Fizzy drinks)

  CPET (oven ready – meal trays)
- Type 2 – HDPE (milk & detergent bottles)
- Type 3 – PVC banned in some countries (water shampoo, Squash etc.)
- Type 4 – LDPE (Plastic bags & bin lines)
- Type 5 – PP (Margarine tubes & microwaveable meal trays)
- Type 6 – PS (Yoghurt pots, plastic cutlery, egg cartons).
- Type 7 – Melamine & non-breakable plates & cup

- ➤ Making of plastic bottles :
- Injection moulding
- Injection blow moulding
- Extrusion blow moulding
- Stretch blow moulding
- Blow moulding
- ➤ **Paper:**
- Grease proof paper
- Kraft paper
- Pouch paper
- Cardboard

Making of paper:

- Pulping method –
  i) Raw material preparation
  ii) Pulping
  iii) Bleaching
  iv) Stock preparation sheet formation
  v) Consolidation
  vi) Drying
  vii) Finishing

Pulping types

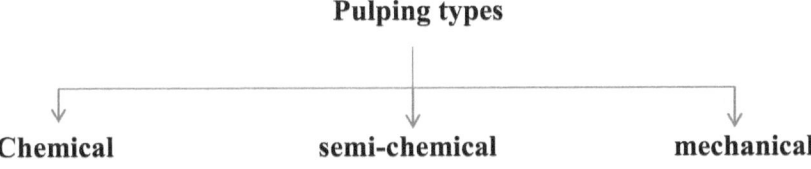

Chemical   **semi-chemical**   **mechanical**

> Bleaching of paper: Chlorine & hypochlorite

Topic 7- Menu

Types of menus:-

1) Table d' hote menu: Means table of the host. This menu is therefore a set menu, in which a no. of dishes are planned by the host & food served & offered at a set price. Selling policy- take it-nor leave it one. Ex- Railways, airlines etc.

 In India in the form of thali meals.

2) A la carte menu: Basically a choice menu and generally offers choice of dishes or items to customers under ten food categories – starters, soups, main dishes, side dishes, cereal preparation, salads, desserts, sandwiches, snacks, beverages.

3) Combination menu: Some establishments it is common to have a la carte menu with a special for the day attachments to it. This special may be a set of dishes, or a plated meal offered in a table a hote form at a set price.

4) Cyclic menu: Hotels, hospitals, homes & other institutions, menus are planned in advance for periods of time varying from five days to one month. These are then cycled or repeated.

Types of menu

Table d'hote	A la carte	Other types
- Banquet	- Breakfast menu	- Static menu
- Buffet	- Lunch menu	- Du Jour menu
- Coffee house	- Dinner menu	- Wine menu
- Cyclical menu	- California menu	- Dessert menu
	- Ethnic menu	
	- Specialty menu	
	- Room service or lounge menu	

1) A la Carte:-

➢ **California menu: -** Where food items are available regardless of the time of the day whether it is breakfast or lunch or dinner can be offered as separate menu.

➢ **Ethnic menu:-** It can be semi a la carte or A'la carte. Offers food items that are representative of the particular cuisine from a particular region or country.

➢ **Lounge menu:-** It offered either in a'la carte or semi a'la carte style. This menus are generally served in hotels, spas. Where customers can order selective item that require easy preparation & less c0stly.

2) Other types:-
- ➢ **Static menu:-** Widely used menus offered by those restaurants those offer same dishes all year long. Ex- fast-food restaurants
- ➢ **Du-Jour menu:-** Other names are 'a plate du jour 'or' specialty of the house' section. This menu usually offers one dish for the particular day which is prepared & changed daily. It also known as " Chalkboard menus"

FOOD SERVICE

Style of service:-

1)

Waiter service		
Banquet service	**Restaurant service**	**Room**
- Formal parties	- Coffee shop	- Hospital
	- Dhabha	- Airline
	- Cafeteria	- Railway
	- Restaurant	

2)

Self service	
Buffet	**Cafeteria**
- Full	- Trayed
- Finger	- Plated
- Fork	

3) Vending:- Seller or vendor bringing food home, or supplying it at bus stations, railway platform, on roadside etc. mobile catering

- Other type of classification of service:-

1) Portable service:- The person has to prepared the food items in her premises & delivered them at the office premises.

2) Self-service:- 4 types

i) Counter

ii) Free flow system:- The counters serve specific items. The customer than goes to the counter from where he selects the items of his choice. Ex- fast food unit pizza, ice-cream, dosa

iii) Buffet:-

iv) Drive-in: The customer is able to pick up the food items from the counter/snacks bar from his vehicle on cash payment. Such a service is seen in drive-in restaurants.

3) Tray service:-

i) Centralized:- Ex- When the food of each patient is place in a tray in the kitchen & delivered to the patient in bed by the use of trolley.

ii) Decentralized:- Ex- private wards. The food from the main kitchen is brought, reheated & assembled in trays & then delivered.

4) Table service:-

i) American style or English style: Customer is greeted by a hostess & the menu card is presented. Waiter takes the orders, places the necessary crockery & cutlery on the table, brings the food from the kitchen, serves the guest & also removed the soiled dishes.

ii) French style: very elaborate, where some preparation or final finishes to the dishes are given on a portable table by the chief waiter & is served assistant waiter. Expensive style

iii) Russian service: Food is completely prepared & portioned in the kitchen.

iv) Banquet service: Any of the style of service i.e. American or Russian is adopted by a large no. of people served together. Ex- conferences & meetings.

v) Gueridon service: "trolley" similar to French style

vi) Filipino service: It is a table service without a waiter/waitress. All courses are served on the table with serving spoon for each course.

vii) Blue plate service: When the group is small, the table is small and the area for dining is small.

Types of Food Service System:-

1) Conventional: Menu item are prepared in a kitchen on place where meals are served & are held a short time, either hot or cold, until they are served. Used by small food service operators.

2) Commissary (Central production kitchen):-

- Is characterized by a large, central production kitchen separate from service units.

- Is employ by airlines, chain restaurants, large school districts.

3) Ready prepared:- Foods are prepared in premises after which they are chilled or frozen & stored for used at some later time. Ex- Hospital & restaurants chains, schools & colleges.

Two ways – i) cook/ chilled method   ii) cook/ freezer method

4) Assembly/serve:- Food production is not done on the premises.

- Fully prepared foods are purchased, stored, assembled, heated & served.

- Used in hospitals & health care institutions.

FOOD COST ANALYSIS

1) Calculation & cost statements:

Percentage Index = Total no. of item A sold / total of all items sold × 100

As = Percentage index of A × Forecast of total customers / 100

Cost involved in catering are-

i) Food cost

ii) Labor cost

iii) Overhead cost

2) Gross profit Ratio:

Gross profit = Total sales – food cost

3) Break even & contribution:- Establishments whose sales figures just cover their variable & fixed costs are said to break-even. When the net profit or loss is zero, the contribution made by the organization just covers the fixed costs.

Quantity:-

Breakeven point = Fixed cost / Average spending power (ASP) – Variable cost per cover

BEP = Fixed cost/ Contribution per cover

Volume of sales:-

BEP = Fixed cost × Selling price/ Selling price × Variable cost per unit

BEP= Fixed cost × Selling price / Contribution per unit

Topic 8- New product Development

New food product: -- It can be defined as the development & introduction of a product not previously manufactured by a company into the market place or the presentation of an old product into a new market not previously explored by the company.

Food product development:-

William J. Stanton:- "Product development encompasses the technical activities of product research, engineering & design".

Need for product development:-

1. Product Decision

2. Size and design of product

3. Name of product

4. Price of product

5. Brand packing & label of product

6. New uses of product

7. Guarantee & after-sale service

Classification & characterization of new food products:-

1. Line Extension: Occur when a company introduces additional items in the same product. Change in flavor, color etc.

Ex- chips – Onion flavor

Tomato flavor

Cream flavor

2. Innovative product: Develop a new innovative product. It helps to company growth, reduce the market risk etc.

3. Creative product

4. New packaging of existing product

5. Reformulation of existing product. Ex- better color, flavor

6. New form of existing product. Ex- solid, powder, concentrated, modified version etc.

7. Repositioned existing product

Product life cycle:-

1. Introduction stage: New product launched in market. It is a stage of slow growth for it. Profits are negative or low because of low sales & high marketing. It may be high chance of success or failure at this stage.

They include 4 strategies→

a) Rapid spamming strategy: This strategy involves introducing the new product with a high promotion & high price in the market.

b) Slow penetration strategy: Low price & low level promotion.

c) Slow spamming strategy: High price & low level promotion. The high price makes much profit at the launching stage of the new product. When market is relatively small in size & customers are well aware of the new product.

d) Rapid penetration strategy: Low price & high level promotion.

2. Growth stage:- New product sales rise at an increasing rate so also there are profitable returns.

3. Maturity stage:- This stage, the sales continue to rise but at a decreasing rate. Need to advertising, sales promotion, encourage consumers to rebuy the product.

4. Saturation stage: - Sales are stable, profits fall drastically & competition is very severe in the market.

5. Decline stage:- Profit is at zero level

Process of development:-

1. Idea generation (of new product): Customers, channel members, employers, competitor's product, research scientists.

2. Screening of new ideas: Pros & cons

3. Concept development & testing

4. Business analysis

5. Market testing

6. Commercialization

www.ingramcontent.com/pod-product-compliance
Lightning Source LLC
Chambersburg PA
CBHW030527220526
45463CB00007B/2750